WHAT'S RENEWABLE ENERGY?

By Jennifer Lombardo

Published in 2022 by
KidHaven Publishing, an Imprint of Greenhaven Publishing, LLC
353 3rd Avenue
Suite 255
New York, NY 10010

Copyright © 2022 KidHaven Publishing, an Imprint of Greenhaven Publishing, LLC.

All rights reserved. No part of this book may be reproduced in any form without permission in writing from the publisher, except by a reviewer.

Designer: Deanna Paternostro
Editor: Jennifer Lombardo

Photo credits: Cover (top) imacoconut/Shutterstock.com; cover (bottom) Maxim Burkovskiy/Shutterstock.com; p. 5 (top) Rasta777/Shutterstock.com; p. 5 (bottom) Africa Studio/Shutterstock.com; p. 7 (main) Minerva Studio/Shutterstock.com; p. 7 (inset) GTW/Shutterstock.com; p. 9 (main) Signature Message/Shutterstock.com; p. 9 (inset) Alf Manciagli/Shutterstock.com; p. 11 esbobeldijk/Shutterstock.com; p. 13 (main) XXLPhoto/Shutterstock.com; p. 13 (inset) Neirfy/Shutterstock.com; p. 15 (main) Encyclopaedia Britannica/UIG via Getty Images; p. 15 (inset) petersemler-photography/Shutterstock.com; p. 17 (main) MicheleB/Shutterstock.com; p. 17 (inset) Puripat Lertpunyaroj/Shutterstock.com; p. 19 (main) Aleks Kend/Shutterstock.com; p. 19 (inset) Alexander Ishchenko/Shutterstock.com; p. 21 best_vector/Shutterstock.com.

Cataloging-in-Publication Data

Names: Lombardo, Jennifer.
Title: What's renewable energy? / Jennifer Lombardo.
Description: New York : KidHaven Publishing, 2022. | Series: What's the issue? | Includes glossary and index.
Identifiers: ISBN 9781534534407 (pbk.) | ISBN 9781534534421 (library bound) | ISBN 9781534534414 (6 pack) | ISBN 9781534534438 (ebook)
Subjects: LCSH: Renewable energy sources–Juvenile literature.
Classification: LCC TJ808.2 L66 2022 | DDC 333.79'4–dc23

Printed in the United States of America

Some of the images in this book illustrate individuals who are models. The depictions do not imply actual situations or events.

CPSIA compliance information: Batch #CS22KH: For further information contact Greenhaven Publishing LLC, New York, New York at 1-844-317-7404.

Please visit our website, www.greenhavenpublishing.com. For a free color catalog of all our high-quality books, call toll free 1-844-317-7404 or fax 1-844-317-7405.

CONTENTS

Getting Power from Nature

Think about all the things you use every day that need power, or energy. Most cars and buses need gasoline, which is made from oil, to run. The lights in your classroom might be powered by a coal **power plant** miles away. The heat in your house might come from a kind of gas called natural gas.

Oil, coal, and gas are types of nonrenewable energy. This means that eventually, we'll use them all up and there won't be any way to quickly get more. To make sure that doesn't happen, people are working on ways to use renewable **resources**, such as the sun, wind, and water, to power the things we need.

Facing the Facts

Another name for coal, oil, and natural gas is "fossil fuels" because they're made from the remains of living things. The oil we use now is made from tiny plants and animals that lived 150 million years ago!

Coal and oil are used to power a lot of things that people use every day. Switching to renewable energy will take a lot of work, and not everyone agrees on the best way to do it.

Warming the Earth

There are many reasons why people want to use renewable resources instead of fossil fuels. We still have enough coal, oil, and gas to use for now. However, while we keep using them, they're causing problems for Earth.

When fossil fuels get into the air, they make it harder for people and animals to breathe. They also trap heat in the **atmosphere**. This makes Earth warmer. When Earth heats up, it causes **climate change**. This makes it harder for plants and animals to live. It also makes storms such as hurricanes, tornadoes, and even snowstorms happen more often.

Facing the Facts

Nine out of the ten warmest years in history happened between 2005 and 2019.

Using fossil fuels changes the climate. This is one reason why people support renewable energy.

Problems for the Land

Oil, gas, and coal all come from deep underground. To get them out, we have to dig deep or use explosives to get rid of the earth and rock above these resources. This changes the land, which can be a big problem for the animals that live there.

When these nonrenewable resources aren't underground anymore, they can get washed into rivers and lakes. This makes the water dirty, so animals can get sick when they drink it or swim in it. People can also get sick when they drink or swim in this dirty water too.

Facing the Facts

In 2019, an oil pipeline in North Dakota was shut down after it leaked 380,000 gallons (1,438,456 L) of oil into the nearby land and water. The amount of oil that leaked would fill about half of an Olympic-size swimming pool!

Oil spills and mines cause problems that could be fixed by using less coal, oil, and gas.

Solar Power

One kind of renewable energy is solar, or sun, power. A special **device** called a solar panel turns sunlight into energy. Solar power doesn't cause climate change the way fossil fuels do, so it's better for the environment, which is the natural world around us. It also won't get used up for millions of years!

Some people put solar panels on their house. These panels cost a lot of money, so not everyone can afford them. Many people think it's better to build solar power plants in very sunny places, such as the desert. Then, the power plants can send power to everyone's homes.

Facing the Facts

In just one second, the sun **releases** more energy than the whole United States uses in one year!

Solar panels such as these work even when it's cloudy, but they don't work at night.

Wind Power

Another kind of renewable energy comes from the wind. A windmill is a machine that turns wind energy into power people can use. Windmills have been used for many years, but the newest ones work better because of new **technology**. This new kind of windmill is called a wind **turbine**.

Wind turbines work best in places without mountains, trees, or tall buildings nearby to block the wind. One good place to put them is near a farm. Another is in the ocean, close to the shore. Like solar power, wind power is safer for the environment than fossil fuels. However, since the wind doesn't blow all the time, it can't be our only energy **source**.

Facing the Facts

If a wind turbine is built in a good spot, it can make enough energy to power about 500 homes each year.

People have been using wind power since 500 CE. Holland is especially famous for its windmills—it has more than 1,000!

Hydropower

In the past, people used a machine called a waterwheel to turn moving water into energy people could use. This is called hydropower. Today, we use water turbines to do this.

One way to use hydropower is to build a dam, which blocks off part of a body of water. When the gates in the dam are raised, the water rushes past the turbine inside. The fast water turns the turbine, making it **generate** power. Raising and lowering the gates controls how much power is generated. Another way to use hydropower is to build a device called a barrage. This works the same as a dam, but it uses the ocean tides to generate power.

Facing the Facts

Hydropower is the kind of energy humans have been using the longest to power machines. People have been using it this way since at least 4000 BCE!

To make water move fast enough to turn a turbine, the water level has to go from high to low. With a barrage, which is shown here, this is controlled by the tide.

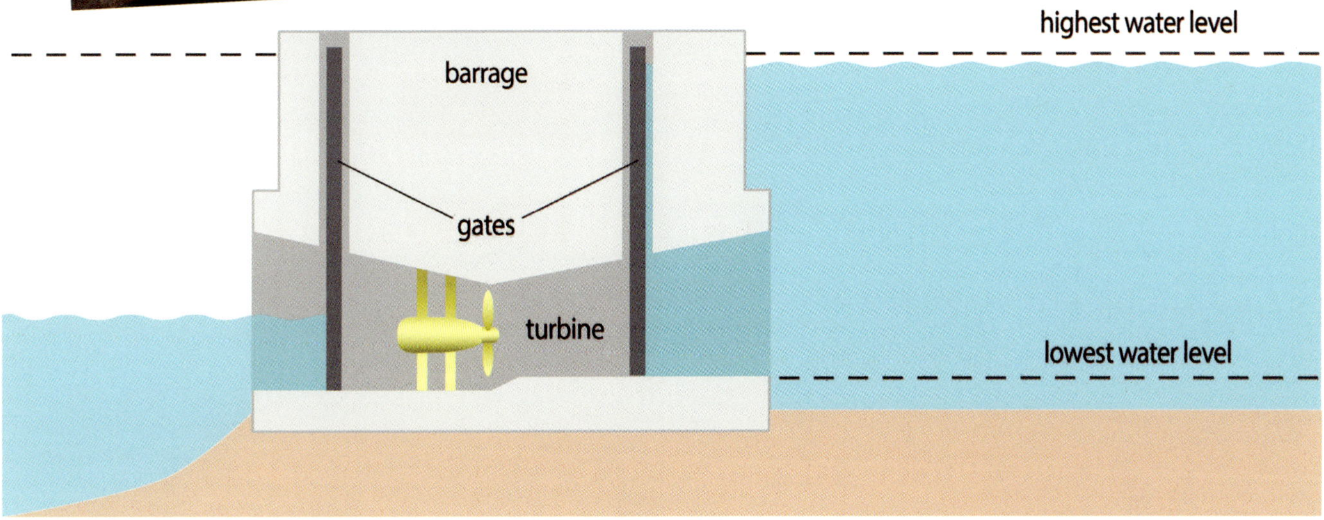

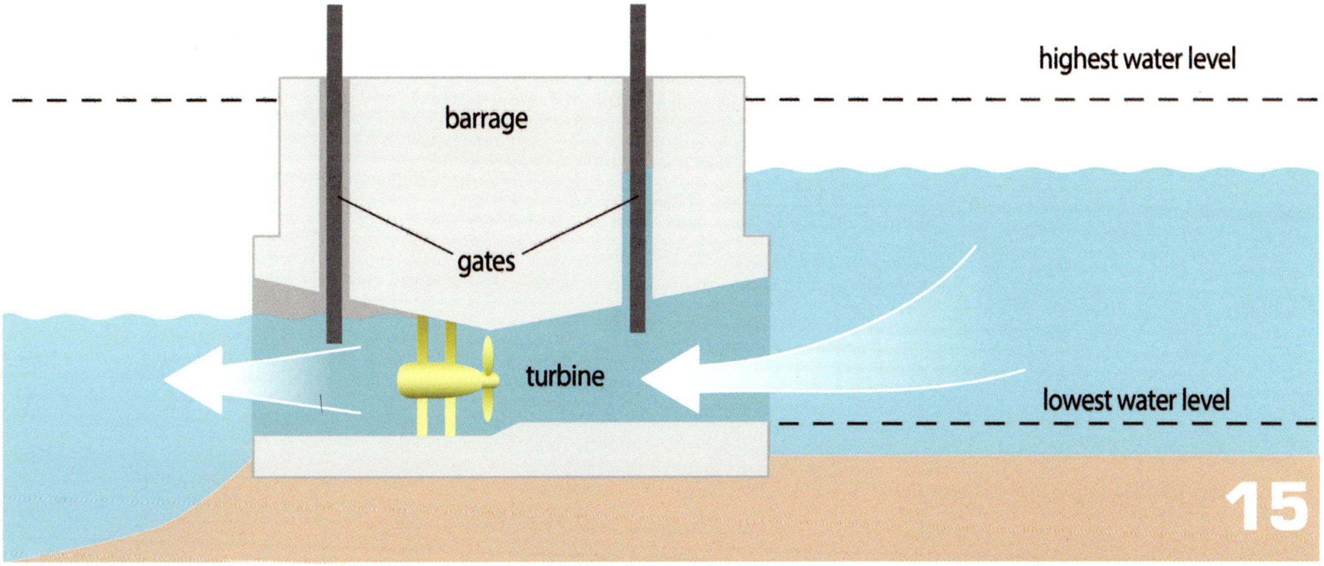

highest water level

barrage

gates

turbine

lowest water level

highest water level

barrage

gates

turbine

lowest water level

Geothermal Power

Geothermal power is heat energy that comes from the earth. "Geo" is a word that means "earth," and "thermal" is a word that means "heat." It's less commonly used than other renewable resources. This is because it can only be used in places where hot water is underground but close to the surface of the earth.

A geothermal power plant takes the heat from this hot water and turns it into energy people can use in their homes. Some people can put a geothermal heating system in their house. This system uses the underground water to heat pipes in their walls. The heat from the pipes warms the house.

Facing the Facts

The largest geothermal power plant in the world is in northern California. It can generate power for more than 22,000 homes!

Iceland has a lot of hot water close to the surface of the earth. This makes it a great place for geothermal power plants.

Biomass

Biomass is a source of energy that comes from plants or animals, which get energy from plants. It's a renewable resource because the energy in plants comes from the sun. Some examples of biomass are certain kinds of trees and grasses; wood chips; corncobs and stalks; and manure, or animal waste.

One way to get energy out of biomass is to burn it. People can use the fire to heat their house or cook their food. Biomass can also be burned with coal. This makes the coal we have last longer and makes coal plants give off less fossil fuel smoke.

Facing the Facts

Biomass is the only renewable resource that can be turned into liquid fuel. This fuel is called ethanol, and it can be added to gasoline to help power cars.

If you've ever made a campfire, you've used biomass energy! A biomass power plant does the same thing, but in a bigger way.

Making a Change

It can be hard to know what to do about renewable energy. Sometimes people feel like there's nothing they can do at all if their local power plant uses nonrenewable resources. However, there are a lot of ways you can use the power of renewable resources at home! The easiest ones to use at your house are sun and wind power.

Now that you know more about renewable resources, which ones do you think would work best where you live? How can people use fossil fuels less often and use renewable resources more often?

Facing the Facts

Experts believe that by 2050, solar power will be the world's main energy source.

WHAT CAN YOU DO?

Turn off things that use electricity, such as lights, when you aren't using them.

Switch to light bulbs that use less energy.

Write to lawmakers. Ask them to pass laws that limit the amount of nonrenewable resources that can be used.

Take shorter showers.

Dry your laundry outside on sunny or windy days.

Go with a trusted adult to a **protest** to let your government leaders know how you feel about using more renewable resources.

These are just some of the things you can do to use less coal, oil, and natural gas and start using more renewable energy sources. Everyone can do their part to help the environment!

GLOSSARY

atmosphere: The air that surrounds Earth.

climate change: Change in Earth's weather caused by human activity.

device: A tool used for a certain purpose.

expert: A person who has special skills or knowledge about something.

generate: To produce something.

power plant: A building in which electric power is created.

protest: An event at which people gather to show strong disapproval about something.

release: To give off.

resource: A supply of something that a person or group has and can use when it's needed.

source: Someone or something that gives what is needed.

technology: A method that uses science to solve problems and the tools used to solve those problems.

turbine: An engine with blades that are caused to spin by pressure from water, steam, or air.

FOR MORE INFORMATION

WEBSITES

BrainPOP: Climate Change

www.brainpop.com/science/earthsystem/climatechange

Through quizzes, movies, and games, this website explains what climate change is and how we can stop it.

Climate Kids

climatekids.nasa.gov/menu/energy

This interactive website explains renewable and nonrenewable energy and how they affect Earth.

BOOKS

Dickmann, Nancy. *Renewable Energy*. London, UK: Wayland, 2019.

Dodd, Emily. *Energy*. London, UK: DK Publishing, 2018.

Sneideman, Joshua, Erin Twamley, and Heather Jane Brinesh. *Renewable Energy: Discover the Fuel of the Future*. White River Junction, VT: Nomad Press, 2016.

Publisher's note to educators and parents: Our editors have carefully reviewed these websites to ensure that they are suitable for students. Many websites change frequently, however, and we cannot guarantee that a site's future contents will continue to meet our high standards of quality and educational value. Be advised that students should be closely supervised whenever they access the Internet.

INDEX